U.S.NRC

United States Nuclear Regulatory Commission

Protecting People and the Environment

NUREG-0090
Vol. 33

Report to Congress on Abnormal Occurrences

Fiscal Year 2010

Office of Nuclear Regulatory Research

AVAILABILITY OF REFERENCE MATERIALS
IN NRC PUBLICATIONS

U.S.NRC

United States Nuclear Regulatory Commission

Protecting People and the Environment

NUREG-0090
Vol. 33

Report to Congress on Abnormal Occurrences

Fiscal Year 2010

Manuscript Completed: June 2011
Date Published: June 2011

Office of Nuclear Regulatory Research

ABSTRACT

Section 208 of the Energy Reorganization Act of 1974 (Public Law 93-438) defines an "abnormal occurrence" (AO) as an unscheduled incident or event that the U.S. Nuclear Regulatory Commission (NRC) determines to be significant from the standpoint of public health or safety. The Federal Reports Elimination and Sunset Act of 1995 (Public Law 104-66) requires that NRC report AOs to Congress annually.

This report describes eight events that NRC identified as AOs during Fiscal Year (FY) 2010 based on the criteria defined in Appendix A to this report. One event involved radiation exposure to an embryo/fetus. The other seven events occurred at NRC-licensed or regulated medical institutions and are medical events as defined in Title 10, Part 35, of the *Code of Federal Regulations* (10 CFR Part 35).

In addition, this report describes seven events that Agreement States identified as AOs during FY 2010 based on the criteria in Appendix A to this report. Agreement States are those States that have entered into formal agreements with NRC, pursuant to Section 274 of the Atomic Energy Act (AEA), to regulate certain quantities of byproduct, source, and special nuclear material at facilities located within their borders. Currently, there are 37 Agreement States. Two Agreement State-licensed events involved radiation exposure to an embryo/fetus. The other five Agreement State-licensed events were medical events as defined in 10 CFR Part 35.

Appendix A to this report presents NRC's criteria for selecting AOs as well as the guidelines for selecting "Other Events of Interest." Appendix B, "Updates of Previously Reported Abnormal Occurrences," provides updated information for two events reported in the FY 2009 "Report to Congress on Abnormal Occurrences." These were medical events at the Gamma Knife Center of the Pacific in Honolulu, Hawaii, and the Veterans Affairs San Diego Health Care System in San Diego, California. During FY 2010, four items were identified as meeting the guidelines for inclusion in Appendix C, "Other Events of Interest." Three of these events occurred at nuclear power plants and the other event occurred at a nuclear fuel cycle facility. Appendix D, "Glossary," contains a glossary of terms used throughout this report. Appendix E, "Conversion Table," presents commonly used conversions when calculating doses.

CONTENTS

Appendices

EXECUTIVE SUMMARY

INTRODUCTION

Section 208 of the Energy Reorganization Act of 1974 (Public Law 93-438) defines an "abnormal occurrence" (AO) as an unscheduled incident or event that the U.S. Nuclear Regulatory Commission (NRC) determines to be significant from the standpoint of public health or safety. The Federal Reports Elimination and Sunset Act of 1995 (Public Law 104-66) requires that NRC report AOs to Congress annually.

This report describes those events that NRC or an Agreement State identified as AOs during Fiscal Year (FY) 2010 based on the criteria defined in Appendix A to this report. Agreement States are those States that have entered into formal agreements with NRC pursuant to Section 274 of the Atomic Energy Act (AEA) to regulate certain quantities of byproduct, source, and special nuclear material at facilities located within their borders. NRC has determined that, of the incidents and events reviewed for this reporting period, only those described herein meet the criteria for being reported as AOs. For each AO, this report documents the date and place, nature, and probable consequences, cause(s), and actions taken to prevent recurrence.

Appendix A to this report presents NRC's criteria for selecting AOs as well as the guidelines for selecting "Other Events of Interest." Appendix B, "Updates of Previously Reported Abnormal Occurrences," provides updated information for two events reported in the FY 2009 "Report to Congress on Abnormal Occurrences." These were medical events at the Gamma Knife Center of the Pacific in Honolulu, Hawaii, and the Veterans Affairs San Diego Health Care System in San Diego, California. During FY 2010, four items were identified as meeting the guidelines for inclusion in Appendix C, "Other Events of Interest." Three of these events occurred at nuclear power plants and the other event occurred at a nuclear fuel cycle facility. Appendix D, "Glossary," contains a glossary of terms used throughout this report. Appendix E, "Conversion Table," presents commonly used conversions when calculating doses.

THE LICENSING AND REGULATORY SYSTEM

The system of licensing and regulation by which NRC carries out its responsibilities is implemented through the rules and regulations in Title 10 of the *Code of Federal Regulations* (10 CFR). Stakeholders are informed and involved, as appropriate, to ensure openness in the agency's regulatory process as stipulated in NRC's Strategic Plan for FY 2008–2013 (NUREG-1614, Volume 4, February 2008). To accomplish its objectives, NRC regularly conducts licensing proceedings, inspection and enforcement activities, operating experience evaluations, and confirmatory research. NRC also maintains programs to establish standards and to issue technical reviews. In addition, NRC considers public participation an essential element of the regulatory process.

NRC adheres to the philosophy that the health and safety of the public are best ensured by establishing multiple levels of protection. These levels are normally achieved and maintained through regulations specifying requirements that ensure the safe use of radioactive materials. Those regulations contain design, operation, and quality assurance criteria appropriate for the various activities regulated by NRC. Licensing, inspection, and enforcement programs provide a regulatory framework to ensure compliance with the regulations. In addition, NRC is striving to make the regulatory system more risk-informed and performance-based, where appropriate.

REPORTABLE EVENTS

NRC initially promulgated the AO criteria in a Commission policy statement published in the *Federal Register* on February 24, 1977 (42 FR 10950), followed by several revisions in subsequent years. The most recent revision to the AO criteria was published in the *Federal Register* on October 12, 2006 (71 FR 60198), and became effective on that date. That revision established the criteria that NRC used to define AOs for the purpose of this report as set forth in Appendix A.

Review and response to operating experience are essential to ensure that licensed activities are conducted safely. Toward that end, the regulations require that licensees must report certain incidents or events to NRC. Such reporting helps to identify deficiencies and ensure that corrective actions are taken to prevent recurrence.

NRC and industry review and evaluate operating experience to identify safety concerns. NRC responds to risk-significant issues through licensing activities and regulations. In addition, the agency maintains operational data in computer-based data files for more effective collection, storage, retrieval, and evaluation.

NRC also routinely disseminates (to the public, industry, and other interested groups) publicly available information and records regarding reportable events at licensed or regulated facilities. The agency achieves this dissemination through public announcements and special notifications to licensees and other affected or interested groups. To widely disseminate information to the public, NRC also issues a *Federal Register* notice describing AOs at facilities (licensed or otherwise regulated by NRC or Agreement States) that occurred in the previous fiscal year. In addition, NRC routinely informs Congress of significant events that occur at licensed or regulated facilities.

AGREEMENT STATES

Section 274 of the AEA, as amended, authorizes the Commission to enter into agreements with States whereby the Commission relinquishes and the States assume regulatory authority over byproduct, source, and special nuclear materials in quantities not sufficient to form a critical mass. States that enter into such agreements with NRC are known as Agreement States. Agreement States must maintain programs that are adequate to protect public health and safety and are compatible with the Commission's program for such materials. At the end of FY 2010, there were 37 Agreement States.

Agreement States report event information to NRC in accordance with compatibility criteria established by the "Policy Statement on Adequacy and Compatibility of Agreement State Programs" that the agency published in the *Federal Register* on September 2, 1997 (62 FR 46517). NRC also has developed and implemented procedures for evaluating materials events to identify those that should be reported as AOs. Toward that end, NRC uniformly applies the AO criteria (in Appendix A to this report) to events at facilities regulated by either NRC or Agreement States. In addition, in early 1977, the Commission determined that the annual report to Congress should include events that meet the criteria for AOs at facilities licensed by Agreement States. Those Agreement State AOs are also included in the *Federal Register* notice that NRC issues to disseminate AO-related information to the public.

FOREIGN INFORMATION

NRC exchanges information with various foreign governments that regulate nuclear facilities. This foreign information is reviewed and considered in NRC's research and regulatory activities as well as in its assessment of operating experience. Although such foreign information may occasionally be referred to in the AO reports to Congress, only domestic AOs are reported.

UPDATES OF PREVIOUSLY REPORTED ABNORMAL OCCURRENCES

NRC provides updates of previously reported AOs if significant new information becomes available. Appendix B, "Updates of Previously Reported Abnormal Occurrences," contains updated information for two events reported in the FY 2009 Report to Congress on Abnormal Occurrences. These were medical events at the Gamma Knife Center of the Pacific in Honolulu, Hawaii, and the Veterans Affairs San Diego Health Care System in San Diego, California.

OTHER EVENTS OF INTEREST

NRC provides information concerning events that are not reportable to Congress as AOs but are included in this report based on the Commission's guidelines as listed in Appendix A. During FY 2010, four items were identified as meeting the guidelines for inclusion in Appendix C, "Other Events of Interest." Three of these events occurred at nuclear power plants and the other event occurred at a nuclear fuel cycle facility.

ABBREVIATIONS

ADAMS	Agencywide Documents Access and Management System
AEA	Atomic Energy Act
AMP	authorized medical physicist
AO	abnormal occurrence
AS	Agreement State
CFR	*Code of Federal Regulations*
cGy	centigray
Ci	curie
cm	centimeter
cm^3	cubic centimeter
CT	computed tomography
DVA	Department of Veterans Affairs
FR	*Federal Register*
FY	Fiscal Year
GBq	gigabecquerel
GSR	gamma stereotactic radiosurgery
Gy	gray
HDR	high dose-rate afterloader
MBq	megabecquerel
mCi	millicurie
ml	milliliter
mm	millimeter
mrem	millirem
mSv	millisievert
No.	number
NOV	Notice of Violation
NRC	U.S. Nuclear Regulatory Commission
OB/GYN	obstetrician/gynecologist
RSO	radiation safety officer
Sv	sievert
TBq	terabecquerel
U.S.	United States

ABNORMAL OCCURRENCES IN FISCAL YEAR 2010

The following is a brief explanation of the outline numbering system used in this section of the report. Appendix A provides the specific criteria for determining when an event is an abnormal occurrence (AO) and provides the guidelines for reporting other events of interest which may not meet the AO criteria but which the Commission has determined should be in this report. Appendix A contains four major categories: I. For All Licensees, II. For Commercial Nuclear Power Plant Licensees, III. Events at Facilities Other Than Nuclear Power Plants and all Transportation Events, and IV. Other Events of Interest. Category IV events are discussed in Appendix C of the report and Categories I, II, and III are discussed in this section. Categories I and II contain significant subelements labeled A, B, C, and D and Category III goes to subelement C. This section of the report only discusses the specific subelement in Categories I, II, and III for which an AO was reported. The identification number for all Agreement State AO reports start with "AS." Similarly, the identification number for all NRC AO reports start with "NRC."

I.   FOR ALL LICENSEES

A.   Human Exposure to Radiation from Licensed Material

During this reporting period, two events at Agreement State-licensed facilities and one event at an NRC-licensed facility were significant enough to be reported as AOs based on criteria in Appendix A to this report. Although these events occurred at medical facilities, they involved unintended exposures to individuals who were not the patient. Therefore, these events belong under criteria I.A, "For All Licensees," as opposed to III.C, "For Medical Licensees."

AS10-01   Human Exposure to Radiation at Mohamed Megahy MD, Ltd in Maryville, Illinois

Criterion I.A.2, "Human Exposure to Radiation from Licensed Material," of Appendix A to this report provides, in part, that any unintended radiation exposure to any minor (an individual less than 18 years of age) resulting in an annual total effective dose equivalent of 50 mSv (5 rem) or more, or to an embryo/fetus resulting in a dose equivalent of 50 mSv (5 rem) or more, shall be considered for reporting as an AO.

Date and Place – May 1, 2007 (reported on June 17, 2010), Maryville, Illinois

Nature and Probable Consequences – Mohamed Megahy MD, Ltd (the licensee) indicated that on May 1, 2007, a patient was given 3,807 MBq (102.9 mCi) of iodine-131 as a treatment for the recurrence of thyroid cancer. On June 11, 2007, the licensee was contacted by the patient's obstetrician/gynecologist (OB/GYN) who advised them that the patient was 25–27 weeks (6 months) pregnant at the time of the iodine-131 administration. At the time of administration, the patient indicated to the licensee that she was not pregnant, and the licensee did not perform an independent test.

In June 2010, the Illinois Emergency Management Agency was contacted by the licensee and requested to make a dose estimate to a fetus as a result of administration of iodine-131 to a patient who was later found to be pregnant. When the Illinois Emergency Management Agency requested additional information to determine the appropriate parameters of the event, the licensee advised the agency that the administration had occurred 3 years earlier. The Illinois Emergency Management Agency calculated an estimated dose to the fetus of 860 mSv (86 rem) and the fetal thyroid of over 1,000,000 mSv (100,000 rem). A full-term child was subsequently born in August 2007 without a thyroid. The child was immediately placed on replacement hormone therapy and continues such treatment.

Cause(s) – The cause of the event was found to be a combination of miscommunication and failure of the licensee to conduct an independent confirmatory pregnancy test.

Actions Taken to Prevent Recurrence

Licensee – The licensee has subsequently made procedural changes to the interview process for screening patients for iodine-131 treatment. This policy includes a confirmatory negative pregnancy test. In addition, the licensee identified the significant delay in reporting the event to the Illinois Emergency Management Agency as not knowing the reporting requirement for this type of event.

State – The Illinois Emergency Management Agency conducted an investigation of the event and issued a Notice of Violation (NOV) for the licensee's failure to report the event. The Illinois Emergency Management Agency is considering rulemaking to require the performance of testing to determine pregnancy prior to administration of iodine-131.

This event is closed for the purpose of this report.

****

AS10-02   **Human Exposure to Radiation at Mercy Medical Center in Durango, Colorado**

Criterion I.A.2, "Human Exposure to Radiation from Licensed Material," of Appendix A to this report provides, in part, that any unintended radiation exposure to any minor (an individual less than 18 years of age) resulting in an annual total effective dose equivalent of 50 mSv (5 rem) or more, or to an embryo/fetus resulting in a dose equivalent of 50 mSv (5 rem) or more, shall be considered for reporting as an AO.

Date and Place – March 16, 2010, Durango, Colorado

Nature and Probable Consequences – Mercy Medical Center (the licensee) reported that a therapeutic dose of 1,110 MBq (30 mCi) of iodine-131 for hyperthyroidism resulted in a dose to an embryo of 80 mGy (8 rem) whole body. Prior to the treatment, the patient informed the licensee's staff that she was not pregnant and the licensee's staff administered a pregnancy test as a routine precaution. The pregnancy test yielded a negative result. Based on the negative pregnancy test results and the patient's interview responses, the licensee administered iodine-131 to the patient.

On April 26, 2010, the patient performed a home pregnancy test that resulted in a positive test result. The patient's pregnancy was confirmed with a positive blood serum pregnancy test on April 27, 2010. The patient's OB/GYN estimated that conception occurred on March 13, 2010 (about 1 week pregnant at the time of administration). A consulting medical physicist reviewed the case and estimated the embryonic exposure (whole body) at 53 to 92 mGy (5.3 to 9.2 rem). The possibility of embryonic thyroid exposure was also investigated and determined to be insignificant due to the early stage of embryonic development. At this dose and administration time in relation to the embryonic development (blastogenesis), the licensee determined that no adverse impact will be likely on subsequent embryonic or fetal development and that subsequent health risks were unlikely. The patient was informed of the dose estimates and potential risks and she elected to continue the pregnancy.

Cause(s) – The cause of this event was the close proximity of conception, which resulted in a negative pregnancy test, to the administration of the iodine-131.

Actions Taken to Prevent Recurrence

Licensee – To help prevent recurrence, the licensee added additional questions to the screening process to help identify patients that might be pregnant even though all procedures to prevent this occurrence were followed.

State – The State conducted an investigation and concurs with the licensee that a reasonable standard of care was met and, consequently, no enforcement action is warranted.

This event is closed for the purpose of this report.

****

NRC10-01   Human Exposure to Radiation at Tripler Army Medical Center in Honolulu, Hawaii

Criterion I.A.2, "Human Exposure to Radiation from Licensed Material," of Appendix A to this report provides, in part, that any unintended radiation exposure to any minor (an individual less than 18 years of age) resulting in an annual total effective dose equivalent of 50 mSv (5 rem) or more, or to an embryo/fetus resulting in a dose equivalent of 50 mSv (5 rem) or more, shall be considered for reporting as an AO.

Date and Place – June 7, 2010, Honolulu, Hawaii

Nature and Probable Consequences – Tripler Army Medical Center (TAMC) (the licensee) reported that a female patient underwent a therapeutic administration of iodine-131 for thyroid ablation therapy. Prior to the treatment, the patient informed the licensee's staff that she was not pregnant and the licensee's staff administered a pregnancy test as a routine precaution. The pregnancy test yielded a negative result. Based on the negative pregnancy test results and the patient's interview responses, the licensee administered iodine-131 to the patient.

On July 8, 2010, the patient became aware that she was pregnant and informed the licensee and her physician. On August 3, 2010, an ultrasound was performed on the patient and a determination was made that the actual date of conception was June 1, 2010 (about 1 week pregnant at time of administration). The TAMC radiation safety officer (RSO) estimated the embryonic dose to be 41.27 cGy (41.27 rad) and concluded that the exposure of the embryo in the first 2 weeks following conception is not likely to result in malformation or embryo/fetal death despite the fact that the central nervous system and the heart are beginning to develop in the third week. NRC contracted with a medical consultant to perform an independent medical evaluation of this embryo/fetal overexposure event. The consultant's report agreed with TAMC conclusions with the exception that the medical consultant did not want to rule out the chance of embryo/fetal malformation.

<u>Cause(s)</u> – The cause of this event was the close proximity of conception, which resulted in a negative pregnancy test, to the administration of the iodine-131.

<u>Actions Taken to Prevent Recurrence</u>

<u>Licensee</u> – The patient consent form has been updated to reflect that the pregnancy test may not show a positive result until the embryo has implanted, which may not occur until 7–10 days after conception. In future consultations, the clinic plans to ask the patient to refrain from any action that may lead to pregnancy during the period immediately prior to therapeutic radioisotope administration.

<u>NRC</u> – NRC conducted an inspection on October 13-14, 2010, and concluded there were no violations of NRC requirements associated with this event.

This event is closed for the purpose of this report.

****

II.   COMMERCIAL NUCLEAR POWER PLANT LICENSEES

During this reporting period, no events at commercial nuclear power plants in the United States were significant enough to be reported as AOs based on the criteria in Appendix A to this report.

****

III. EVENTS AT FACILITIES OTHER THAN NUCLEAR POWER PLANTS AND ALL TRANSPORTATION EVENTS

C. Medical Licensees

During this reporting period, seven events at NRC-licensed or regulated facilities and five events at Agreement State-licensed facilities were significant enough to be reported as AOs based on criteria in Appendix A to this report.

AS10-03   Medical Event at Mercy St. Vincent Medical Center in Toledo, Ohio

Criterion III.C.1.b, III.C.2.a and III.C.2.b.(iii), "For Medical Licensees," of Appendix A to this report provides, in part, that a medical event shall be considered for reporting as an AO if it results in a dose equal to or greater than 10 Gy (1,000 rad) to any organ or tissue (other than a major portion of the bone marrow, or the lens of the eye, or the gonads); and represents either a dose or dosage that is at least 50 percent greater than that prescribed; or is a prescribed dose delivered to the wrong treatment site.

Date and Place – November 8, 2005 (reported on March 3, 2010), Toledo, Ohio

Nature and Probable Consequences – Mercy St. Vincent Medical Center (the licensee) reported that a medical event occurred associated with a brachytherapy seed implant procedure to treat prostate cancer. The patient was prescribed to receive a total dose of 160 Gy (16,000 rad) to the prostate using 67 iodine-125 seeds. Instead, the patient's sigmoid colon received at least the full prescription dose of 160 Gy (16,000 rad) and a significant portion of the bladder base including the region of the urethral orifices received at least 108 Gy (10,800 rad) (wrong treatment sites). The patient and referring physician were informed of this event.

On March 3, 2010, the Ohio Department of Health (ODH) performed an inspection of the licensee and noted that the licensee had not reported this medical event to the State and the NRC. The licensee had not identified the medical event as a reportable event and did not investigate it to determine a cause. Subsequently, the licensee reported the medical event to the NRC. The licensee confirmed that 13 of the permanent iodine-125 seeds were improperly positioned in the bladder and subsequently removed from the patient's bladder immediately after the procedure. A post-implant dose calculation showed that the prostate received a dose of 15.43 Gy (1,543 rad), or 9.6 percent of the prescribed dose. The patient chose to then receive an external beam treatment with a linear accelerator to treat the tumor. About 13 months after the brachytherapy procedure, the patient developed rectosigmoid bleeding that required hospitalization and argon laser coagulopathy. In August 2010, ODH ordered an independent medical expert evaluation of the event. The independent medical expert concluded that the subsequent delivery of external beam radiotherapy may have contributed to the rectosigmoid damage, but the high dose from the brachytherapy procedure almost certainly was the primary cause of the damage.

Cause(s) – The cause of the medical event was the failure of the licensee to adequately visualize the prostate prior to the implant procedure.

Actions Taken To Prevent Recurrence

Licensee – Corrective actions taken by the licensee included training of the RSO, medical physicist, clinical director, and radiation oncologists on ODH regulations concerning medical events. New procedures were also developed for brachytherapy seed implant procedures.

State – In March 2010, ODH conducted a special inspection of the licensee and issued an NOV. The NOV required the licensee to perform a self audit of all brachytherapy cases performed since November 2004, which revealed seven additional medical events that were not reported. In June 2010, an Adjudication Order and administrative penalty of $25,000 were issued to the licensee.

The event is closed for the purpose of this report.

NRC10-02   Medical Event at Chippenham & Johnston-Willis (CJW) Medical Center in Richmond, Virginia

Criterion III.C.1.b and III.C.2.b.(iii), "For Medical Licensees," of Appendix A to this report provide, in part, that a medical event shall be considered for reporting as an AO if it results in a dose equal to or greater than 10 Gy (1,000 rad) to any organ or tissue (other than a major portion of the bone marrow, or the lens of the eye, or the gonads) and represents a prescribed dose or dosage that is delivered to the wrong treatment site.

Date and Place – December 16, 2008, Richmond, Virginia

Nature and Probable Consequences – Chippenham & Johnston-Willis (CJW) Medical Center (the licensee) reported a medical event with its gamma stereotactic radiosurgery (GSR) unit. A patient being treated for trigeminal neuralgia (inflammation of the nerve) was prescribed a treatment of 40 Gy (4,000 rad) to the right trigeminal nerve but received the treatment dose to the left trigeminal nerve (wrong treatment site). The patient and referring physician were informed of this event.

The licensee noted that on the day of the treatment, the top portion of the written directive correctly documented the prescribed treatment site; however, while the staff was preparing the daily patient treatment log, it was inadvertently annotated that the dose was to be delivered to the left trigeminal nerve. This error was carried through by the medical physicist during preparation of the patient's treatment plan and completion of the bottom part of the written directive. Upon completion of the procedure and after reviewing the patient's file, the treatment team identified the inadvertent treatment of the left trigeminal nerve. The NRC contracted medical consultant concluded that although no actual consequences resulted, an unlikely injury to the brain stem was possible due to high radiation dose to a tiny volume of the brain stem tissue and an increased risk of cataract formation.

Cause(s) – The cause of the medical event was the licensee's failure to have adequate procedures that verify the location of treatment sites and ensure that any inconsistencies in the written directives are resolved prior to administration.

Actions Taken to Prevent Recurrence

Licensee – The licensee revised their GSR treatment procedures to affirm that (1) a "Physician Order" will be the primary source of documentation of the treatment site and will accompany the patient through the entire course of the treatment, (2) the radiation oncologist and the neurosurgeon will independently verify and document the treatment site, (3) the nurse and the medical physicist will confirm that the treatment site identified by the radiation oncologist in the written directive and the neurosurgeon's "physician order" both match, (4) the neurosurgeon will mark the treatment site with ink in the presence of a nurse, and (5) a "Time-Out" process involving independent verification of the final treatment plan by each of the four members of the clinical team (who are required to sign-off their presence and acceptance of time-out in the presence of the patient before moving ahead with the treatment) will be used with the patient or the patient's authorized representative to confirm the treatment site.

NRC – NRC initiated an inspection on December 18, 2008. NRC completed the inspection on November 30, 2009, and issued one Severity Level III violation to the licensee on January 21, 2010.

The event is closed for the purpose of this report.

NRC10-03   Medical Event at Virtua Health System in Marlton, New Jersey

Criterion III.C.1.b and III.C.2.b.(iii), "For Medical Licensees," of Appendix A to this report provide, in part, that a medical event shall be considered for reporting as an AO if it results in a dose equal to or greater than 10 Gy (1,000 rad) to any organ or tissue (other than a major portion of the bone marrow, or the lens of the eye, or the gonads) and represents a prescribed dose or dosage that is delivered to the wrong treatment site.

Date and Place – January 19, 2009, Marlton, New Jersey

Nature and Probable Consequences – Virtua Health System (the licensee) reported that a medical event occurred associated with a brachytherapy seed implant procedure to treat prostate cancer. The patient was prescribed to receive a total dose of 145 Gy (14,500 rad) to the prostate using 93 iodine-125 seeds. Instead, the patient received an approximate dose of 12.2 Gy (1,220 rad) to the rectum (wrong treatment site). The patient and referring physician were informed of this event.

On January 19, 2009, the urologist inserted needles in the patient's prostate gland under transrectal ultrasound guidance while the radiation oncologist left the operating room to obtain the radioactive seeds. The licensee's staff (including the authorized medical physicist [AMP]) questioned the accuracy of prostate visualization prior to implantation of the seeds but took no action to resolve the question. On February 23, 2009, following a post-implant computed tomography (CT) scan, it was noted that some mispositioning of the sources occurred and the patient was notified that additional treatment may be necessary. On March 19, 2009, the AMP reviewed the case and determined that 100 percent of the seeds were implanted outside of the prostate, which received about 10 Gy (1,000 rad). NRC contracted with a medical consultant who concluded that although the probability of long-lasting negative health effects to the patient is low an increased risk of impotency and fibrosis was possible due to the high radiation dose.

Cause(s) – The cause of the medical event was failure of the medical implant team to adequately visualize and identify the prostate prior to the implant.

Actions Taken To Prevent Recurrence

Licensee – The licensee revised its policy and procedures to require that (1) all members of the implant team be present before the patient is brought to the operating room and placed under anesthesia, (2) the AMP be included in the pre-implantation ultrasound, (3) the authorized user consult with the urologist before needle insertion, (4) both the radiation oncologist and the urologist agree on the positioning and the visualizing of the target anatomy, (5) any objection or question by an implant team member is cause for stopping the implant and performing a review, and (6) the implant be stopped if there are any ultrasound image questions. The licensee's staff was also trained on the revised procedures, the definition and reporting requirements of a medical event, and the communication of any CT scan abnormalities or seed misplacement to the RSO.

<u>NRC</u> – NRC initiated an inspection on March 20, 2009. NRC completed the inspection on August 26, 2009, and issued one Severity Level III violation to the licensee on October 21, 2009.

This event is closed for the purpose of this report.

NRC10-04   Medical Event at Nanticoke Memorial Hospital, in Seaford, Delaware

Criterion III.C.1.b and III.C.2.b.(iii), "For Medical Licensees," of Appendix A to this report provide, in part, that a medical event shall be considered for reporting as an AO if it results in a dose equal to or greater than 10 Gy (1,000 rad) to any organ or tissue (other than a major portion of the bone marrow, or the lens of the eye, or the gonads) and represents a prescribed dose or dosage that is delivered to the wrong treatment site.

Date and Place – March 5, 2009 (reported on July 15, 2009), Seaford, Delaware

Nature and Probable Consequences – Nanticoke Memorial Hospital (the licensee) reported that a medical event occurred involving a brachytherapy seed implant procedure to treat prostate cancer. The patient was prescribed a total dose of 145 Gy (14,500 rad) to the prostate using 61 iodine-125 seeds. Instead, the patient received an approximate prostate dose of 26 Gy (2,600 rad) (18 percent of the prescribed dose) and a dose of 139 Gy (13,900 rad) to unintended tissue (wrong treatment site). The patient and referring physician were informed of this event.

The seeds were implanted under ultrasound guidance using an axial view; however, following the implant, the urologist performed a cystoscopy to remove 22 of the seeds from the bladder. When the patient returned to the hospital for a post-implant CT scan, the images revealed that 32 seeds were displaced superiorly to the prostate and 7 seeds were implanted in the prostate. NRC contracted with a medical consultant who concluded that no significant adverse health effects to the patient were expected.

Cause(s) - The cause of the medical event was due to a miscalculation of the prostate depth in relation to the skin surface due to possible patient movement during the procedure.

Actions Taken to Prevent Recurrence

Licensee - The licensee revised its prostate implant procedure to include the use of both the axial and sagittal views of an ultrasound probe to determine prostate depth. In addition, the licensee revised its medical event policy to ensure timely reporting of medical events and to clearly state the parameters under which a medical event must be reported. The licensee provided training on the revised policies and procedures to its staff.

NRC – NRC initiated an inspection on July 19, 2009. NRC completed the inspection on January 6, 2010, and issued one Severity Level III violation to the licensee on February 2, 2010.

The event is closed for the purpose of this report.

AS10-04   Medical Event at Hoag Memorial Hospital Presbyterian in Newport Beach, California

Criterion III.C.1.b and III.C.2.b.(iii), "For Medical Licensees," of Appendix A to this report provide, in part, that a medical event shall be considered for reporting as an AO if it results in a dose equal to or greater than 10 Gy (1,000 rad) to any organ or tissue (other than a major portion of the bone marrow, or the lens of the eye, or the gonads) and represents a prescribed dose or dosage that is delivered to the wrong treatment site.

Date and Place – March 20, 2009, Newport Beach, California

Nature and Probable Consequences – Hoag Memorial Hospital Presbyterian (the licensee) reported that a medical event occurred associated with its GSR unit. A patient being treated for an acoustic neuroma was scheduled to receive between 11 and 18 Gy (1,100 and 1,800 rads) to an intended neuroma volume of 0.08 cm^3 but, due to an unintended shift in the treatment volume of about 2 mm, only about one-half of the neuroma received the treatment dose and an adjacent temporal bone volume of 0.04 cm^3 received the treatment dose (wrong treatment site). The other half of the neuroma received between 3 and 11 Gy (300 and 1,100 rads). The patient and physician were informed of this event.

The unintended shift in treatment volume occurred due to a misaligned fiduciary marker (indicator) box during a CT scan used in the treatment planning process. The misalignment occurred because one alignment pin of four on the indicator box was not fully seated in the stereotactic frame attached to the patient's head, resulting in the indicator box not being correctly aligned. The alignment pin error was not detected until the conclusion of the treatment. The additional dose to the temporal bone because of the alignment error is not expected to result in any significant adverse health effect to the patient.

Cause(s) – The medical event is believed to have been caused by human error in not ensuring the CT indicator box was properly installed at the time of the CT scan. It is not known if the improper installation occurred when the technologist positioned the indicator box in the stereotactic frame or whether the indicator box became misaligned during patient positioning in preparation for the CT scan.

Actions Taken to Prevent Recurrence

Licensee – The licensee has retrained all CT technologists concerning the proper placement of the CT indicator box. Also, because use of CT imaging for GSR treatment is infrequent (normally MRI is used), the licensee now requires that a GSR qualified medical physicist verify the placement of the CT indicator box immediately prior to all CT imaging that will be used for GSR treatment planning.

State – On June 22, 2009, the California Department of Public Health (CDPH) issued an NOV related to this event. Subsequently, CDPH received dosimetry information which they used to interpret the event as not meeting the AO criteria; however, CDPH was not certain of this determination and asked NRC for a final determination. On July 1, 2010, after the NRC Medical Radiation Safety Team (MSRT) had performed a careful analysis of the event along with the dosimetry data, NRC determined that the event met the AO criteria.

This event is closed for the purpose of this report.

AS10-05   Medical Event at Marshfield Clinic in Marshfield, Wisconsin

Criterion III.C.1.b and III.C.2.a, "For Medical Licensees," of Appendix A to this report provide, in part, that a medical event shall be considered for reporting as an AO if it results in a dose equal to or greater than 10 Gy (1,000 rad) to any organ or tissue (other than a major portion of the bone marrow, or the lens of the eye, or the gonads) and represents a dose or dosage that is at least 50 percent greater than that prescribed.

Date and Place – June 2005 to May 2007, (reported on July 8, 2010) Marshfield, Wisconsin

Nature and Probable Consequences – In July 2010, the Marshfield Clinic (the licensee) reviewed all prostate brachytherapy cases performed under its license in the past 7 years. The review resulted in the identification of nine medical events involving permanent implants of iodine-125 for prostate brachytherapy where the total dose delivered differed from the prescribed dose by 20 percent or more, or another organ received at least 50 percent more dose than intended. The three medical events involved planned doses to the prostate of 120 Gy (12,000 rad), 160 Gy (16,000 rad), and 160 Gy (16,000 rad). The licensee assumes an identical planned dose to the urethra. However, these treatments resulted in actual doses to the urethra of 191.6 Gy (19,160 rad), 258.1 Gy (25,810 rad), and 242.6 Gy (24,260 rad), which were overdoses of 59.7, 61.3, and 51.6 percent, respectively. The licensee notified the affected patients and referring physicians.

The authorized user physicians had previously determined that patients would not suffer significant health effects for urethral doses below 400 Gy (40,000 rad). Because the urethra penetrates through the center of the prostate and the prostate itself is a small gland, a balance exists between reducing the dose to the urethra and delivering the prescribed dose to the prostate. The doses delivered to the patients in question were well within the 400 Gy (40,000 rad) urethral tolerance dose, and the licensee considered the treatments to be clinically acceptable.

Cause(s) – The licensee suspects that the implants deviated from their intended tracks after insertion into the prostate, causing the seeds to be deposited closer to the urethra.

Actions Taken to Prevent Recurrence

Licensee – Corrective actions included developing a procedure for ensuring that treatments were delivered in accordance with the written directive, planning treatments to D90 (minimum dose received by 90 percent of CT-defined prostate volume) values of 100–110 percent, using the same written directive form at each site that performs brachytherapy, increasing ultrasound and fluoroscopy visualization during prostate implants and providing additional training to personnel.

State – The Wisconsin Department of Health Services determined that Marshfield Clinic did not have a procedure for evaluating whether the dose delivered in a prostate brachytherapy treatment was in accordance with the written directive. In addition, the licensee did not have criteria for identifying a medical event for prostate brachytherapy. The licensee has been cited for several items of noncompliance.

This event is closed for the purpose of this report.

NRC10-05   Medical Event at Yale New-Haven Hospital, in New Haven, Connecticut

Criterion III.C.1.b and III.C.2.b.(iii), "For Medical Licensees," of Appendix A to this report provide, in part, that a medical event shall be considered for reporting as an AO if it results in a dose equal to or greater than 10 Gy (1,000 rad) to any organ or tissue (other than a major portion of the bone marrow, or the lens of the eye, or the gonads) and represents a prescribed dose or dosage that is delivered to the wrong treatment site.

Date and Place – August 5, 2009, New Haven, Connecticut

Nature and Probable Consequences – Yale New-Haven Hospital (the licensee) reported that a medical event occurred associated with its GSR unit. A patient being treated for brain metastases was prescribed 18 Gy (1,800 rad). However, while treating a patient earlier in the day, an equipment malfunction occurred with the GSR unit that resulted in a positioning shift of the x-axis by 4.5 mm. The positioning shift in the x-axis resulted in an underdose to the treatment site and an overdose to a wrong treatment site. The patient and physician were informed of this event.

The malfunction occurred following the treatment of the first patient on August 5, 2009. The automatic positioning system (APS) malfunctioned and, after discussion with the GSR manufacturer, the position error codes were cleared by the AMP. A second patient was treated for multiple brain metastases later that day. GSR service personnel noted on August 5, 2009, that the APS positioning was off by about 5 mm. After further evaluation, the manufacturer determined that a position shift (offset) occurred when licensee personnel accepted an error message concerning position deviation. NRC contracted with a medical consultant who concluded that no clinically significant side effects from radiation damage to the wrong treatment sites would be expected.

Cause(s) – The cause of the medical event was failure of licensee personnel to verify that the APS coordinates were in accordance with the written directive.

Actions Taken to Prevent Recurrence

Licensee – The licensee issued a memorandum to all personnel involved in GSR treatments to require visual verification of the physical coordinates against the electronic coordinates before the start and at the end of each treatment run. The licensee also retrained all GSR personnel on the importance of fully understanding error conditions and reviewing unexpected errors with other staff involved in the treatment (e.g., radiation oncologist, AMP, etc.) prior to clearing any unexpected error.

NRC – NRC initiated an inspection on August 13, 2009. NRC completed the inspection on April 7, 2010, and issued one Severity Level III violation to the licensee on May 21, 2010.

The event is closed for the purpose of this report.

NRC10-06   Medical Event at Valley Hospital in Paramus, New Jersey

Criterion III.C.1.b and III.C.2.b.(iii), "For Medical Licensees," of Appendix A to this report provide, in part, that a medical event shall be considered for reporting as an AO if it results in a dose equal to or greater than 10 Gy (1,000 rad) to any organ or tissue (other than a major portion of the bone marrow, or the lens of the eye, or the gonads) and represents a prescribed dose or dosage that is delivered to the wrong treatment site.

Date and Place – July 29, 2009, Paramus, New Jersey

Nature and Probable Consequences – Valley Hospital (the licensee) reported that a medical event occurred associated with a brachytherapy seed implant procedure to treat prostate cancer. The patient was prescribed a total dose of 65 Gy (6,500 rad) to the prostate using 46 cesium-131 seeds. Instead, the licensee determined that an unintended volume (30.1 ml) of soft tissue received 100 percent of the prescribed prostate dose. The patient and referring physician were informed of this event.

On August 6, 2009, the patient returned to the hospital for a post-implant CT scan. The images revealed that the seeds were implanted in soft tissue 4 to 5 cm from the prostate. Post-implant dosimetry calculations indicated that none of the prostate received the prescribed dose of 6,500 cGy (6,500 rad). NRC contracted with a medical consultant who concluded that the additional dose can increase the risk of soft tissue fibrosis or increase the risk of impotency.

Cause(s) – The cause of the medical event was the licensee's failure to identify the position of the prostate due to the patient's unusual anatomy and obesity.

Actions Taken To Prevent Recurrence

Licensee – The licensee revised their prostate implant procedures to include steps to ensure that the prostate and surrounding anatomy is adequately visualized prior to implant.

NRC – NRC initiated an inspection on August 13, 2009. NRC completed the inspection on October 29, 2009, and determined that no violations of NRC requirements occurred.

This event is closed for the purpose of this report.

NRC10-07　Medical Event at Christiana Care Health Center in Wilmington, Delaware

Criterion III.C.1.b and III.C.2.b.(iii), "For Medical Licensees," of Appendix A to this report provides, in part, that a medical event shall be considered for reporting as an AO if it results in a dose equal to or greater than 10 Gy (1,000 rad) to any organ or tissue (other than a major portion of the bone marrow, or the lens of the eye, or the gonads) and represents either a dose or dosage that is delivered to the wrong treatment site.

Date and Place – January 18, 2010, Wilmington, Delaware

Nature and Probable Consequences – Christiana Care Heath Center (the licensee) reported that a patient was prescribed a high dose-rate (HDR) mammosite (brachytherapy) multi-lumen catheter treatment of 34 Gy (3,400 rad) over a 5 day period to the left breast. The patient received an average dose of 17 Gy (1,700 rad) to 100 cm^3 of unintended breast tissue; 68 Gy (6,800 rad) to 7.5 cm^3 of unintended skin and underlying tissue; and 3.4 Gy (340 rad) to 35 cm^3 of intended breast tissue. The patient and referring physician were informed of this event.

On February 22, 2010, during a follow-up examination, the patient complained about skin reddening on the external breast. In reviewing the treatment plan, it was discovered that the AMP performed measurements using a source position simulator (SPS) measurement tool following a CT scan to determine the treatment distance for each catheter. The catheter distances were recorded and confirmed with two manufacturer representatives that were present at the time of the treatment. However, it was noted that an incorrect measurement caused the placement of the radioactive source 10 cm proximal to the intended position. The NRC-contracted medical consultant concluded that the dose that was administered to the unintended left breast tissue is unlikely to result in any significant or unusual adverse effect. However, a significant risk exists that local tumor recurrence could occur if additional intervention is not performed.

Cause(s) – The cause of the medical event was human error in the failure to identify that the measurement tool was functioning improperly and to identify an incorrect measurement distance.

Actions Taken To Prevent Recurrence

Licensee – The licensee revised its procedures for HDR brachytherapy to require a double-check of all patient measurements, a daily and monthly quality assurance requirement to confirm that the SPS tool is functioning properly, and a process to ensure that all members of the treatment team agree on the specifics of the treatment. In addition, the licensee acquired a new SPS tool, developed and posted a reference table at the HDR control console, provided training on revised procedures to staff involved in the HDR program (to be repeated annually), and implemented a "New Product" committee to review all new product plans.

NRC – NRC conducted an inspection on July 12, 2010, and issued one Severity Level III violation to the licensee on August 24, 2010.

The event is closed for the purpose of this report.

AS10-06   Medical Event at Mary Bird Perkins Cancer Center in Baton Rouge, Louisiana

Criterion III.C.1.b, and III.C.2.b.(iii), "For Medical Licensees," of Appendix A to this report provides, in part, that a medical event shall be considered for reporting as an AO if it results in a dose equal to or greater than 10 Gy (1,000 rad) to any organ or tissue (other than a major portion of the bone marrow, or the lens of the eye, or the gonads) and is a prescribed dose delivered to the wrong treatment site.

Date and Place – March 15, 2010, Baton Rouge, Louisiana

Nature and Probable Consequences – Mary Bird Perkins Cancer Center (the licensee) reported that a medical event occurred associated with a brachytherapy seed implant procedure to treat prostate cancer. The patient was prescribed a total dose of 145 Gy (14,500 rad) to the prostate using iodine-125 seeds. Instead, the patient received a dose of 39.55 Gy (3,955 rad) to the rectum, 40.94 Gy (4,094 rad) to the urethra, and 6 Gy (600 rad) to the bladder (wrong treatment sites). The patient and referring physician were informed of this event.

During the review of this event, the licensee determined that a positioning error occurred and the dose was delivered about 3.0 cm away from the targeted prostate gland. The estimated dose to the prostate gland was 12.88 Gy (1,288 rad). The licensee concluded that no significant adverse health effect to the patient is expected.

Cause(s) – The medical event was caused by patient movement between the time the planning images were obtained and the actual implantation of the seeds.

Actions Taken To Prevent Recurrence

Licensee – The licensee modified its procedure to insert the needles that hold the prostate in place prior to obtaining the ultrasound images instead of immediately before the seed needles are inserted. In addition, the sagittal image will be captured at the time of planning image acquisition and confirmed periodically throughout the case, and the radiation oncologist will personally confirm the location of the reference base prior to dispensing the first seed.

State – The Louisiana Department of Environmental Quality conducted an investigation, reviewed the licensee's corrective actions, and found the corrective actions to be adequate.

This event is closed for the purpose of this report.

AS10-07   Medical Event at Mayo Clinic in Rochester, Minnesota

Criterion III.C.1.b, III.C.2.a and III.C.2.b.(iii), "For Medical Licensees," of Appendix A to this report provides, in part, that a medical event shall be considered for reporting as an AO if it results in a dose equal to or greater than 10 Gy (1,000 rad) to any organ or tissue (other than a major portion of the bone marrow, or the lens of the eye, or the gonads); represents either a dose or dosage that is at least 50 percent greater than that prescribed; and is a prescribed dose delivered to the wrong treatment site.

<u>Date and Place</u> – March 23, 2010, Rochester, Minnesota

<u>Nature and Probable Consequences</u> – The Mayo Clinic (the licensee) reported a medical event associated with an HDR biliary treatment for liver carcinoma containing 329 GBq (8.9 Ci) of iridium-192. A patient was prescribed to receive four fractionated doses totaling 16 Gy (1,600 rad) to the liver. The treatment to the liver should have produced an estimated dose to the duodenum (wrong treatment site) of 1.2 Gy (120 rad) but as a result of the event it received a dose of about 10 Gy (1,000 rad). The patient and referring physician were informed of this event.

During the second fractioned treatment, the measurement cable was inserted into the catheter and it was noted that it extended about 17 cm beyond the programmed treatment distance used during the first fractioned treatment. It was concluded that the measurement wire on the first treatment had met with some resistance at a tight bend and that it was not at the end of the catheter. This resulted in overdosing the duodenum (wrong treatment site). Upon discovery of the treatment distance error and overdose, the licensee changed the written directive to add a fifth fractioned treatment to correct for the underdose of the liver. A lesser total dose to the liver was given because of concerns regarding the dose already received by the duodenum. The authorized user concluded that no chronic health effect to the patient is expected.

<u>Cause(s)</u> – The medical event was caused by human error in failing to verify that the correct catheter length was entered into the HDR unit.

<u>Actions Taken To Prevent Recurrence</u>

<u>Licensee</u> – The licensee committed to taking several corrective actions including the imaging of inserted catheters prior to treatments and performing catheter length checks prior to HDR treatments.

<u>State</u> – On April 6, 2010, the Minnesota Department of Health (MDH) staff performed a reactive inspection of the licensee's HDR program. The MDH approved the licensee's corrective actions and did not take enforcement action.

This event is closed for the purpose of this report.

NRC10-08   Medical Event at Providence Hospital in Novi, Michigan

Criterion III.C.1.b and III.C.2.b.(iii), "For Medical Licensees," of Appendix A to this report provide, in part, that a medical event shall be considered for reporting as an AO if it results in a dose equal to or greater than 10 Gy (1,000 rad) to any organ or tissue (other than a major portion of the bone marrow, or the lens of the eye, or the gonads) and represents a prescribed dose or dosage that is delivered to the wrong treatment site.

Date and Place – August 30, 2010, Novi, Michigan

Nature and Probable Consequences – Providence Hospital (the licensee) reported that a medical event occurred associated with an anal brachytherapy treatment using 32 seeds containing iodine-125. The intended dose was 90 Gy (9,000 rad) to the tumor. Instead, the patient's seminal vesicle received 19.79 Gy (1,979 rad) more than intended and the bladder received 3.68 Gy (368 rad) more than intended. The patient and referring physician were informed of this event.

On September 1, 2010, a follow-up CT scan showed that the permanent implants had been inserted about 4 cm from the intended location. The licensee reported that the tumor near the anus and rectum received a maximum dose of 8 Gy (800 rad). The licensee calculated the dose difference to the surrounding tissue as a result of the improper permanent implant placement. The licensee concluded that no significant adverse health effect to the patient is expected.

Cause(s) – The licensee determined that the cause of the event was that they did not use tissue markers to confirm source placement and the insertion needle did not have a visible mark to ensure proper depth placement.

Actions Taken to Prevent Recurrence

Licensee – Procedures were modified to administer sources as prescribed in the written directive as follows: (1) any interstitial procedure that requires the use of fluoroscopy alone will be done with the use of tissue markers to confirm source placement, and (2) interstitial procedures that use fluoroscopy alone will have needle depth verified. The licensee completed training of licensee staff on the event and the corrective actions by October 1, 2010.

NRC – Region III reviewed and concurred on the licensee's corrective actions. NRC has retained the services of an independent medical consultant to determine if any significant health effects to the patient are expected.

This event is open for the purpose of this report.

APPENDIX A
ABNORMAL OCCURRENCE CRITERIA AND
GUIDELINES FOR OTHER EVENTS OF INTEREST

An accident or event will be considered an AO if it involves a major reduction in the degree of protection of public health or safety. This type of incident or event would have a moderate or more severe impact on public health or safety and could include, but need not be limited to, the following:

(1)   Moderate exposure to, or release of, radioactive material licensed by or otherwise regulated by the Commission;

(2)   Major degradation of essential safety-related equipment; or

(3)   Major deficiencies in design, construction, use of, or management controls for facilities or radioactive material licensed by or otherwise regulated by the Commission

The following criteria for determining an AO and the guidelines for "Other Events of Interest" were stated in an NRC policy statement published in the *Federal Register* on October 12, 2006 (71 FR 60198).

Abnormal Occurrence Criteria

Criteria by types of events used to determine which events will be considered for reporting as AOs are as follows:

I.   For All Licensees

      A.   Human Exposure to Radiation from Licensed Material

           1.   Any unintended radiation exposure to an adult (any individual 18 years of age or older) resulting in an annual total effective dose equivalent (TEDE) of 250 mSv (25 rem) or more; or an annual sum of the deep dose equivalent (external dose) and committed dose equivalent (intake of radioactive material) to any individual organ other than the lens of the eye, the bone marrow, and the gonads of 2,500 mSv (250 rem) or more; or an annual dose equivalent to the lens of the eye of 1 Sv (100 rem) or more; or an annual sum of the deep dose equivalent and committed dose equivalent to the bone marrow of 1 Sv (100 rem) or more; or a committed dose equivalent to the gonads of 2,500 mSv (250 rem) or more; or an annual shallow-dose equivalent to the skin or extremities of 2,500 mSv (250 rem) or more.

           2.   Any unintended radiation exposure to any minor (an individual less than 18 years of age) resulting in an annual TEDE of 50 mSv (5 rem) or more, or to an embryo/fetus resulting in a dose equivalent of 50 mSv (5 rem) or more.

           3.   Any radiation exposure that has resulted in unintended permanent functional damage to an organ or a physiological system as determined by a physician.

B. Discharge or dispersal of radioactive material from its intended place of confinement which results in the release of radioactive material to an unrestricted area in concentrations which, if averaged over a period of 24 hours, exceeds 5,000 times the values specified in Table 2 of Appendix B to Title 10 of the *Code of Federal Regulations* (10 CFR) Part 20, unless the licensee has demonstrated compliance with §20.1301 using §20.1302(b)(1) or §20.1302(b)(2)(ii). This criterion does not apply to transportation events.

C. Theft, Diversion, or Loss of Licensed Material, or Sabotage or Security Breach[1,2]

    1. Any unrecovered lost, stolen, or abandoned sources that exceed the values listed in Appendix P to Part 110, "High Risk Radioactive Material, Category 2." Excluded from reporting under this criterion are those events involving sources that are lost, stolen, or abandoned under the following conditions: sources abandoned in accordance with the requirements of 10 CFR 39.77(c); sealed sources contained in labeled, rugged source housings; recovered sources with sufficient indication that doses in excess of the reporting thresholds specified in AO criteria I.A.1 and I.A.2 did not occur while the source was missing; and unrecoverable sources (sources that have been lost and for which a reasonable attempt at recovery has been made without success) lost under such conditions that doses in excess of the reporting thresholds specified in AO criteria I.A.1 and I.A.2 are not known to have occurred and the agency has determined that the risk of theft or diversion is acceptably low.

    2. A substantiated[3] case of actual theft or diversion of licensed, risk-significant radioactive sources or a formula quantity[4] of special nuclear material; or act that results in radiological sabotage[5].

    3. Any substantiated[3] loss of a formula quantity[4] of special nuclear material or a substantiated[3] inventory discrepancy of a formula quantity[4] of special nuclear material that is judged to be caused by theft or diversion or by a substantial breakdown[6] of the accountability system.

    4. Any substantial breakdown[6] of physical security or material control (i.e., access control containment or accountability systems) that significantly weakened the protection against theft, diversion, or sabotage.

    5. Any significant unauthorized disclosures (loss, theft, and/or deliberate) of classified information that harms national security or safeguards information that harms the public health and safety.

D. Initiation of High-Level NRC Team Inspections.[7]

II. For Commercial Nuclear Power Plant Licensees

A. Malfunction of Facility, Structures, or Equipment

    1. Exceeding a safety limit of license technical specification (TS) [10 CFR 50.36(c)].

2.   Serious degradation of fuel integrity, primary coolant pressure boundary, or primary containment boundary.

3.   Loss of plant capability to perform essential safety functions so that a release of radioactive materials which could result in exceeding the dose limits of 10 CFR Part 100 or 5 times the dose limits of 10 CFR Part 50, Appendix A, General Design Criterion (GDC) 19, could occur from a postulated transient or accident (e.g., loss of emergency core cooling system, loss of control rod system).

B.   Design or Safety Analysis Deficiency, Personnel Error, or Procedural or Administrative Inadequacy

    1.   Discovery of a major condition not specifically considered in the safety analysis report or TS that requires immediate remedial action.

    2.   Personnel error or procedural deficiencies that result in loss of plant capability to perform essential safety functions so that a release of radioactive materials which could result in exceeding the dose limits of 10 CFR Part 100 or 5 times the dose limits of 10 CFR Part 50, Appendix A, GDC 19, could occur from a postulated transient or accident (e.g., loss of emergency core cooling system, loss of control rod drive mechanism).

C.   Any reactor events or conditions that are determined to be of high safety significance.[8]

D.   Any operating reactor plants that are determined to have overall unacceptable performance or that are in a shutdown condition as a result of significant performance problems and/or operational event(s).[9]

III.   Events at Facilities Other than Nuclear Power Plants and all Transportation Events

A.   Events Involving Design, Analysis, Construction, Testing, Operation, Transport, Use, or Disposal of Licensed Facilities or Regulated Materials

    1.   An accidental criticality [10 CFR 70.52(a)].

    2.   A major deficiency in design, construction, control, or operation having significant safety implications that require immediate remedial action.

    3.   A serious safety-significant deficiency in management or procedural controls.

    4.   A series of events (in which the individual events are not of major importance), recurring incidents, or incidents with implications for similar facilities (generic incidents) that raise a major safety concern.

B.	For Fuel Cycle Facilities

1.	Absence or failure of all safety-related or security-related controls (engineered and human) for an NRC-regulated lethal hazard (radiological or chemical) while the lethal hazard is present.

2.	An NRC-ordered safety-related or security-related immediate remedial action.

C.	For Medical Licensees

A medical event that:

1.	Results in a dose that is
	a.	Equal to or greater than 1 Gy (100 rad) to a major portion of the bone marrow or to the lens of the eye; or equal or greater than 2.5 Gy (250 rad) to the gonads; or

	b.	Equal to or greater than 10 Gy (1,000 rad) to any other organ or tissue; and

2.	Represents either
	a.	A dose or dosage that is at least 50 percent greater than that prescribed, or
	b.	A prescribed dose or dosage that
		(i)	Uses the wrong radiopharmaceutical or unsealed byproduct material; or
		(ii)	Is delivered by the wrong route of administration; or
		(iii)	Is delivered to the wrong treatment site; or
		(iv)	Is delivered by the wrong treatment mode; or
		(v)	Is from a leaking source or sources; or
		(vi)	Is delivered to the wrong individual or human research subject.

IV. Other Events of Interest

The Commission may determine that events other than AOs may be of interest to Congress and the public and should be included in an appendix to the AO report as "Other Events of Interest." Such events may include, but are not necessarily limited to, events that do not meet the AO criteria but that have been perceived by Congress or the public to be of high health and safety significance, have received significant media coverage, or have caused the NRC to increase its attention to or oversight of a program area, or a group of similar events that have resulted in licensed materials entering the public domain in an uncontrolled manner.

---

[1] Information pertaining to certain incidents may be either classified or under consideration for classification because of national security implications. Classified information will be withheld when formally reporting these incidents in accordance with Section 208 of the ERA of 1974, as amended. Any classified details regarding these incidents would be available to the Congress, upon request, under appropriate security arrangements.

[2] Due to increased terrorist activities worldwide, the AO report would not disclose specific classified information and sensitive information, the details of which are considered useful to a potential terrorist. Classified information is defined as information that would harm national security if disclosed in an unauthorized manner.

[3] "Substantiated" means a situation where an indication of loss, theft, or unlawful diversion such as: an allegation of diversion, report of lost or stolen material, statistical processing difference, or other indication of loss of material control or accountability cannot be refuted following an investigation; and requires further action on the part of the Agency or other proper authorities.

[4] A formula quantity of special nuclear material is defined in 10 CFR 70.4.

[5] Radiological sabotage is defined in 10 CFR 73.2.

[6] A substantial breakdown is defined as a red finding in the security inspection program, or any plant or facility determined to have overall unacceptable performance, or in a shutdown condition (inimical to the effective functioning of the Nation's critical infrastructure) as a result of significant performance problems and/or operational events.

[7] This subelement addresses initiation of any Incident Investigation Teams, as described in NRC Management Directive (MD) 8.3, "NRC Incident Investigation Program," or initiation of any Accident Review Groups, as described in MD 8.9, "Accident Investigation."

[8] The NRC ROP uses four colors to describe the safety significance of licensee performance. As defined in NRC Management Directive 8.13, "Reactor Oversight Process," green is used for very low safety significance, white is used for low to moderate safety significance, yellow is used for substantial safety significance, and red is used for high safety significance. Reactor conditions or performance indicators evaluated to be red are considered Abnormal Occurrences. Additionally, Criterion II.C also includes any events or conditions evaluated by the NRC ASP program to have a conditional core damage probability (CCDP) or change in core damage probability (ΔCDP) of greater than 1×10^{-3}.

[9] Any plants assessed by the ROP to be in the unacceptable performance column, as described in NRC Inspection Manual Chapter 0305, "Operating Reactor Assessment Program." This assessment of safety performance is based on the number and significance of NRC inspection findings and licensee performance indicators.

APPENDIX B
UPDATES OF PREVIOUSLY REPORTED ABNORMAL OCCURRENCES

During this reporting period, updated information became available for two AO events the NRC previously reported in the "Report to Congress on Abnormal Occurrences: Fiscal Year (FY) 2009" regarding the medical events at the Gamma Knife Center and the Department of Veterans Affairs.

Medical Event at the Gamma Knife Center (previously reported as NRC09-02 in NUREG-0090, Volume 32)

Date and Place – July 2, 2009, Gamma Knife Center of the Pacific in Honolulu, Hawaii

Background – The Gamma Knife Center of the Pacific reported that a medical event occurred associated with its GSR unit. A patient being treated for multiple brain metastatic sites received a dose of 24 Gy (2,400 rad) to additional brain tissue. The cause of the additional dose was the erroneous use of a collimator helmet containing orifices that were larger in diameter than prescribed. The licensee concluded that no significant adverse health effect to the patient was expected. Corrective actions taken by the licensee included sending a notice to all authorized users, neurosurgeons, and medical physicists reiterating that they should each independently check the collimator size prior to patient treatment and revising procedures to have a second independent verification of all treatment parameters including the collimator size, by a treatment team member. NRC conducted an onsite inspection and hired a medical consultant to review the event. The full details of the event are discussed in the FY 2009 abnormal occurrence report as NRC09-02.

Update on Actions Taken To Prevent Recurrence – NRC issued a NOV for the licensee's failure to have written procedural requirements that demonstrate with a high degree of confidence that an administration is in accordance with the treatment plan and the written directive. NRC did not pursue imposition of a civil penalty because the licensee had not been the subject of escalated enforcement actions within the last two inspections and implemented prompt and comprehensive corrective actions. In July 2010, the licensee replaced its GSR unit with a newer design that does not rely on manual change out of collimator helmets and reported that the patient did not show any health effects attributable to the use of the incorrect collimator size.

This event is closed for the purpose of this report.

****

Medical Event at the Veterans Affairs San Diego Health Care System (previously reported as NRC09-03 in NUREG-0090, Volume 32)

Date and Place – September 21, 2009, Veterans Affairs San Diego Health Care System in San Diego, California

Background – The Department of Veterans Affairs (the licensee), National Health Physics Program reported that a medical event occurred at the Veterans Affairs San Diego Health Care System associated with a therapeutic dosage of iodine-131 for the treatment of metastatic thyroid cancer. A patient was prescribed to receive 6.9 GBq (187 mCi) of iodine-131 to the metastatic sites around the body but received 6.1 GBq (166 mCi) to the stomach (wrong treatment site). The patient and the referring physician were informed of this event. The root causes of the event were identified as (1) inadequate procedures, (2) inadequate training of staff personnel, and (3) an inadequate verification process of written directives involving administrations with gastric tubes. Corrective actions taken by the licensee included suspension of one individual's participation in administrations requiring a written directive, informal training of the nuclear medicine technologists by the RSO, and the development of draft written policies and procedures on the administration of iodine-131 through a gastric tube. NRC conducted a reactive inspection in November 2009 and hired a medical consultant to review the event. The full details of the event are discussed in the FY 2009 AO report as NRC09-03.

Update on Actions Taken To Prevent Recurrence – The NRC medical consultant did not identify any adverse health effects to the patient as a result of the unintended dose to the patient's stomach. In addition, on June 2, 2010, NRC issued a NOV and Proposed Imposition of Civil Penalty to the Department of Veterans Affairs in the amount of $14,000.

This event is closed for the purpose of this report.

****

APPENDIX C
OTHER EVENTS OF INTEREST

This appendix discusses "Other Events of Interest" that do not meet the AO criteria in Appendix A but have been perceived by Congress or the public to be of high health and safety significance, have received significant media coverage, or have caused NRC to increase its attention to or oversight of a program area, including a group of similar events that have resulted in licensed materials entering the public domain in an uncontrolled manner.

EOI-01   Three Mile Island Nuclear Power Plant: Low-Level Contamination Event with Media Interest

This event is being included in this report because it received significant media attention and was perceived by the public as well as the national and international media to be of high health and safety significance. However, as described below the actual radiation exposure to the affected workers was less than 1 percent of NRC regulatory limits and this event was actually of low safety significance.

Exelon (the licensee) reported that, on November 22, 2009, vacuuming was being conducted in a steam generator to remove debris that could potentially impact decontamination equipment. The workers did not recognize that the vacuum cleaner lacked a high-efficiency particulate air (HEPA) filter and that it was dispersing airborne radioactivity into the containment building causing alarms on various monitors. No radioactivity was detectable above background outside the containment construction opening.

The alarming monitors prompted a containment evacuation of about 175 workers. Of the 175 workers, 145 workers were determined to have either low-level external radioactive contamination or low-level intakes of airborne radioactivity associated with the event. Workers were evacuated from containment in a timely manner (about 27 minutes) and a further evaluation of internal uptakes of radiation reduced the total number of those contaminated.

The levels of contamination were low and did not pose a health or safety concern to the workers or public. The maximum radiation dose to any worker from this event was less than 0.2 mSv (20 mrem). The annual Federal limit for nuclear plant worker exposure is 50 mSv (5,000 mrem); thus the maximum occupational doses due to this event were less than 1 percent of this NRC regulatory limit. To put these levels in perspective, the average American receives about 6 mSv (600 mrem) of radiation exposure each year from natural background sources, such as cosmic, terrestrial and internal radiation, as well as from nuclear medicine procedures and treatments. Another useful reference measure is the amount of radiation in our bodies from the food and water we ingest (such as naturally occurring radioactive potassium-40), which is estimated at 0.4 mSv (40 mrem) per year, or more than twice the exposure received by any worker during this event.

NRC follow-up inspections determined that no issues were identified associated with operational "reactor safety." However, three findings of very low safety significance (green findings) were identified and documented as noncited violations and entered into the licensee's corrective action program. The NRC inspectors independently evaluated Exelon's radiological assessment relative to public health and safety and confirmed that the offsite environmental releases during the event were within regulatory limits. Exelon appropriately documented the evaluation of releases from the containment in its corrective action program. Exelon also collected and analyzed river water samples and downwind owner-controlled area soil samples. The NRC inspectors' review indicated that no radioactivity was detected in the samples that were attributable to this event at Three Mile Island Unit 1. The follow-up inspection of this event is fully documented in NRC Inspection Report: Three Mile Island Station, Unit 1 NRC Inspection Report 5000289/2010007 available at ADAMS Accession No. ML101050517.

****

EOI-02   Nuclear Power Plants: Leaks in Underground Pipes, Groundwater Contamination and Tritium Issues

This item is being included in this report because tritium leaks in underground pipes and groundwater contamination issues at nuclear power plants have received significant public, media, and Congressional interest. Tritium is a mildly radioactive isotope of hydrogen that occurs both naturally and during the operation of nuclear power plants. Nuclear plants normally release authorized radioactive effluents under NRC effluent discharge limitations including water containing tritium. The leaks of tritium to groundwater are typically a very small fraction of the authorized radioactive effluents that are discharged to surface water. The pipe degradation leading to these leaks has not affected the operability of safety systems.

Over the past 30 years, instances of buried piping leaks have occurred in safety-related and non-safety-related piping at about 50 percent of the nuclear power plant sites. These tritium leaks have caused localized groundwater contamination that has not contaminated drinking water wells. Generally, groundwater flows down-gradient and offsite into a large water body such as a river or lakes where normal discharges of radioactive effluents occur. These leaks have not resulted in exceeding any public health and safety standards or exceeding any operational controls that are used to keep radioactive effluents as low as is reasonably achievable. In fact, no drinking water has been affected; however, public interest has been expressed concerning the impact of these leaks on environmental resources such as drinking water supplies.

In March 2010, NRC's Executive Director of Operations (EDO) established a Groundwater Task Force (GTF) to review NRC's approach to overseeing buried pipes given the recent incidents of leaking buried pipes at commercial nuclear power plants. The charter of the GTF was to reevaluate the recommendations made in the Liquid Radioactive Release Lessons Learned Task Force Final Report dated September 1, 2006; review the actions taken in the Commission paper SECY-09-0174 (Staff Progress in Evaluation of Buried Piping at Nuclear Reactor Facilities, ADAMS Accession No. ML093160004); and review the actions taken in response to recent releases of tritium into groundwater by nuclear facilities.

The GTF completed its work in June 2010 and provided its report to the EDO. The report characterized a variety of issues ranging from policy issues to communications improvement opportunities. The complete report may be found under ADAMS Accession No. ML101740509. The GTF determined that NRC is accomplishing its stated mission of protecting public health, safety, and protection of the environment through its response to groundwater leaks/spills. Within the current regulatory structure, NRC is correctly applying requirements and properly characterizing the relevant issues. However, the GTF reported that further observations, conclusions, and recommendations exist that NRC should consider in its oversight of licensed materials outside of its design confinement.

The EDO appointed a Senior Management Review Group (SMRG) to consider the findings of the GTF report and determine how to address the conclusions and recommendations in the final report. The SMRG identified both near-term staff actions and potential policy issues for consideration. The SMRG assessment of the first two themes of the GTF report was addressed in SECY Paper 2011-0019. The complete paper may be found under ADAMS Accession No. ML110050525. The second two themes of the GTF report are addressed in a memorandum to the Chairman titled "Initiatives for Improved Communication of Groundwater Incidents," (ADAMS Accession No. ML110050252).

NRC has held public meetings and engaged the public in response to these concerns. In addition, NRC has established Web pages to keep the public informed of the ongoing developments related to these issues. For more details and additional web page links see the NRC web page: "Buried Reactor Pipes and Tritium" available at: http://www.nrc.gov/reading-rm/doc-collections/fact-sheets/buried-pipes-tritium.html.

****

EOI-03   H.B. Robinson Nuclear Power Plant: Event Resulting in an Augmented Inspection

On March 28, 2010, an electrical fault and fire occurred at the H.B. Robinson Steam Electric Plant Unit 2, resulting in a reactor trip and subsequent safety injection actuation due to a rapid cooldown of the reactor coolant system. In addition, the reactor coolant pump (RCP) seals experienced a concurrent loss of seal injection and thermal barrier heat exchanger cooling. During the event response, operators successfully restored cooling water to the thermal barrier heat exchanger prior to seal failure.

During normal operation, the charging system supplies seal injection water to the RCP pump seals. The seal injection water cools the RCP seals and bearings, and also prevents hot reactor coolant from entering the seals and bearing areas. The purpose of the thermal barrier heat exchanger is to cool any reactor coolant leaking up the shaft to protect the radial bearing and shaft seals. The component cooling water system provides the source of cooling for this heat exchanger. A loss of both of these systems can lead to overheating of the RCP seals and pump bearings with subsequent failure of the RCP seals that could result in leakage from the primary coolant system.

Approximately 4 hours after the event began and after the plant was placed in a stable shutdown state, operators inadvertently reinitiated the electrical fault and fire, causing further damage to surrounding equipment. Due to the additional equipment damage, the licensee declared an Alert emergency classification. The Alert was terminated in the early morning of March 29, 2010.

On June 2, 2010, the NRC completed an augmented inspection that identified 14 unresolved issues. The analysis of these issues revealed five findings of very low safety significance (green) and two findings of low to moderate safety significance (white). The two white findings involved operators failing to implement proper command and control and the licensee failing to correctly implement proper training protocols in their Licensed Operator Requalification Program. The Final Determination Letter and Notice of Violation regarding these two white findings were issued in a letter to the licensee dated January 31, 2011 (NRC Inspection Report No. 05000261/2011008, available at ADAMS Accession No. ML110310469). Evaluation of these findings revealed that crew experience and composition, main control room ergonomics, and reliance upon knowledge-based emergency operating procedures were complicating factors during the event response. The licensee has implemented corrective actions that include, but are not limited to, enhancements to licensed operator training material, re-training and evaluation of all control room operators, procedure enhancements, crew reconstitution to enhance performance, and personnel and management changes. Additionally, the NRC has performed inspections to verify that important operational safety aspects have been addressed.

This event does not currently meet the AO reporting criteria; however, recent information identified during NRC supplemental inspection activities could potentially cause the NRC to determine that this event is of high safety significance (Criterion II.C). At this time staff is evaluating the event under the NRC's Accident Sequence Precursor (ASP) program.

The ASP Program provides an integrated risk analysis of all deficiencies, equipment failures, and degraded conditions that were observed during the event. The inspection program separately assesses the risk associated with each performance deficiency. Therefore, for events involving multiple licensee performance deficiencies and equipment failures, as in the H.B. Robinson event, it is not unexpected that the ASP and inspection programs would assign different risk significance levels. As such, the integrated approach used by the ASP Program complements the inspection program. In the case of the H.B. Robinson event, the staff has concluded that the preliminary results from the integrated ASP analysis are consistent with the risk significance of the two white inspection findings.

If the final ASP analysis of this event results in its identification as a significant precursor, the NRC will report this event in Section II, "Commercial Nuclear Power Plant Licensees," of next fiscal year's AO Report and in the FY 2011 Performance and Accountability Report to Congress.

****

FACILITIES OTHER THAN NUCLEAR POWER PLANTS

EOI-04   Nuclear Fuel Services Inc.: Adverse Chemical Reaction Event

This event is the result of an adverse chemical reaction that did not result in a release of radioactivity but is included in this report because it caused NRC to increase its attention and oversight to this program area.

On October 13, 2009, Nuclear Fuels Services (NFS) (the licensee) experienced an unexpected exothermic chemical reaction within the Blended Low Enriched Uranium Preparation Facility. The elevated temperatures from the reaction created nitrogen compound gases within the associated process off-gas piping. An instrument located near the ceiling of the facility detected these gases and generated an alarm that resulted in the evacuation of employees from the affected area. In addition, the elevated temperature of these gases caused portions of the plastic off-gas piping system to deform and sag. NFS personnel took action to shut down the system and as a result, no personnel were injured and offsite environmental releases during the event were within regulatory limits.

In response to the event, NRC formed a Special Inspection Team that arrived at the licensee's facility on October 19, 2009. NRC upgraded its response to an Augmented Inspection Team following notification by the licensee of their analysis of the event. The licensee's analysis revealed that, based on the specific type of material processed in the event, the nitrogen compound gases generated could have resulted in high occupational consequences. As defined in Title 10, Part 70, Section 61 (b)(4) of the *Code of Federal Regulations* (10 CFR Part 70.61 (b)(4)), high occupational consequences refers to an acute chemical exposure to an individual from hazardous chemicals produced from licensed material.

The preliminary results of the augmented inspection and an interim review of the licensee's overall safety performance identified a number of concerns regarding the licensee's ability to provide reasonable assurance of its ability to safely operate the facility. These concerns involved the adequacy of the licensee's management oversight of facility process changes, perceived production pressures, lack of questioning attitude by workers and management, and poor communications. In addition, NRC identified concerns with the decisions made by the licensee's management in both October and November 2009 to restart the uranium aluminum process lines without fully understanding the causes of the events and without correcting the underlying problems.

On January 7, 2010, NRC issued a Confirmatory Action Letter regarding commitments made by the licensee in a letter dated December 30, 2009. The actions included (1) suspending operation of several processing lines, (2) completing specific actions before restart of operations, and (3) providing NRC with sufficient time to inspect completion of the actions. After extensive team inspections, NRC authorized the restart of four processing lines in March 2010, May 2010, July 2010, and October 2010 respectively. Portions of one process line remain shutdown pending equipment modifications and restart inspections.

On September 2, 2010, NRC imposed a civil penalty of $140,000 based on a Severity Level III problem involving three violations associated with the event. The penalty was paid in October 2010. The three violations involved (1) failure to have adequate engineered or administrative controls for operations in violation of 10 CFR 70.61(b), (2) failure to comply with multiple facility operating procedures regarding the facility system change process, and (3) failure to maintain records necessary to support the licensee's determination that specific facility changes did not require prior NRC approval in violation of 10 CFR 70.72.

APPENDIX D
GLOSSARY

Absorbed Dose – as defined in 10 CFR 20.1003, the energy imparted by ionizing radiation per unit mass of irradiated material; the units of absorbed dose are the rad and the gray (Gy).

[I]**Acoustic neuroma** – a nonmalignant usually slow-growing tumor involving the Schwann cells of a vestibular nerve that may cause deafness, tinnitus, and disturbance of the sense of balance and may be life threatening if not treated.

Act – as defined in 10 CFR 40.4, the Atomic Energy Act of 1954 (68 Stat. 919) including any amendments thereto.

Authorized User (AU) – as defined in 10 CFR 35.2, a physician who (1) meets the requirements in §§35.59 and 35.190(a), 35.290(a), 35.390(a), 35.392(a), 35.394(a), 35.490(a), 35.590(a), or 35.690(a); or (2) is identified as an authorized user on (i) a Commission or Agreement State license that authorizes the medical use of byproduct material; (ii) a permit issued by a Commission master material licensee that is authorized to permit the medical use of byproduct material; (iii) a permit issued by a Commission or Agreement State specific licensee of broad scope that is authorized to permit the medical use of byproduct material; or (iv) a permit issued by a Commission master material license broad scope permittee that is authorized to permit the medical use of byproduct material.

[I]**Biliary** — of, relating to, or conveying bile.

[I]**Blastogenesis** – the transformation of lymphocytes into larger cells capable of undergoing mitosis.

Brachytherapy – as defined in 10 CFR 35.2, a method of radiation therapy in which sources are used to deliver a radiation dose at a distance of up to a few centimeters by surface, intracavitary, intraluminal, or interstitial application.

Brachytherapy Source – as defined in 10 CFR 35.2, a radioactive source or a manufacturer-assembled source train or a combination of these sources that is designed to deliver a therapeutic dose within a distance of a few centimeters.

[I]**Catheter** – a tubular medical device for insertion into canals, vessels, passageways, or body cavities for diagnostic or therapeutic purposes to permit injection or withdrawal of fluids or to keep a passage open.

[I]**Coagulopathy** – a disease or condition affecting the blood's ability to coagulate.

[I]**Computed Tomography (CT)** – radiography in which a three-dimensional image of a body structure is constructed by computer from a series of cross-sectional images made along an axis.

[II]**Cystoscopy** – a procedure in which the doctor inserts a lighted instrument called a cystoscope into the urethra in order to look inside the urethra and bladder.

Dose Equivalent (H$_T$) – as defined in 10 CFR 20.1003, the product of the absorbed dose in tissue, quality factor, and all other necessary modifying factors at the location of interest; the units of dose equivalent are the rem and sievert.

"Duodenum – the first, shortest, and widest part of the small intestine that in humans is about 10 inches (25 centimeters) long and extends from the pylorus to the undersurface of the liver where it descends for a variable distance and receives the bile and pancreatic ducts and then bends to the left and finally upward to join the jejunum near the second lumbar vertebra.

Effective Dose Equivalent (H_E) – as defined in 10 CFR 20.1003, the sum of the products of the dose equivalent to the organ or tissue (H_T) and the weighting factors (w_T) applicable to each of the body organs or tissues that are irradiated ($H_E = \Sigma\, w_T\, H_T$).

Exposure – as defined in 10 CFR 20.1003, being exposed to ionizing radiation or to radioactive material.

External Dose – as defined in 10 CFR 20.1003, that portion of the dose equivalent received from radiation sources outside the body.

¹Fibrosis – a condition marked by increase of interstitial fibrous tissue.

¹Gamma Knife – a type of radiosurgery (radiation therapy) machine that acts by focusing low-dosage gamma radiation from many sources on a precise target. Areas adjacent to the target receive only slight doses of radiation while the target gets the full intensity. The gamma knife may be used to treat brain tumors, meningiomas (tumors on the protective layers of the brain), and trigeminal neuralgia causing severe facial pain.

Gray (Gy) – as defined in 10 CFR 20.1004, the international system of unit of absorbed dose; one gray is equal to an absorbed dose of 1 Joule/kilogram (100 rads).

High Dose-Rate (HDR) Remote Afterloader – as defined in 10 CFR 35.2, a brachytherapy device that remotely delivers a dose rate in excess of 12 Gy (1,200 rad) per hour at the point of surface where the dose is prescribed.

¹Interstitial – situated within but not restricted to or characteristic of a particular organ or tissue, used especially of fibrous tissue.

¹Lumen – the bore of a tube (as of a hollow needle or catheter).

"Mammosite Treatment – a minimally invasive radiation therapy technique used to treat breast cancer. This technique uses brachytherapy to deliver radiation directly to the site of the tumor bed from inside the body. A soft balloon, attached to a thin catheter, is inserted into the cavity where the tumor was removed. The balloon is inflated, and a computer-controlled machine delivers the radiation down the catheter into the balloon where it irradiates the tumor bed.

Manual Brachytherapy – as defined in 10 CFR 35.2, a type of brachytherapy in which the brachytherapy sources (e.g., seeds, ribbons) are manually placed topically on or inserted either into the body cavities that are in close proximity to a treatment site or directly into the tissue volume.

Medical Event – as defined in 10 CFR 35.2, an event that meets the criteria in §35.3045(a) or (b). 10 CFR 35.3045(a) states that a licensee shall report any event, except for an event that results from patient intervention, in which the administration of byproduct material or radiation from byproduct material results in (1) a dose that differs from the prescribed dose or dose that would have resulted from the prescribed dosage by more than 0.05 Sv (5 rem) effective dose equivalent, 0.5 Sv (50 rem) to an organ or tissue, or 0.5 Sv (50 rem) shallow dose equivalent to the skin and (i) the total dose delivered differs from the prescribed dose by 20 percent or more, (ii) the total dosage delivered differs from the prescribed dosage by 20 percent or more or falls outside the prescribed dosage range, or (iii) the fractionated dose delivered differs from the prescribed dose for a single fraction by 50 percent or more; (2) a dose that exceeds 0.05 Sv (5 rem) effective dose equivalent, 0.5 Sv (50 rem) to an organ or tissue, or 0.5 Sv (50 rem) shallow dose equivalent to the skin from any of the following (i) an administration of a wrong radioactive drug containing byproduct material, (ii) an administration of a radioactive drug containing byproduct material by the wrong route of administration, (iii) an administration of a dose or dosage to the wrong individual or human research subject, (iv) an administration of a dose or dosage delivered by the wrong mode of treatment, or (v) a leaking sealed source; (3) a dose to the skin or an organ or tissue other than the treatment site that exceeds by 0.5 Sv (50 rem) to an organ or tissue and 50 percent or more of the dose expected from the administration defined in the written directive (excluding, for permanent implants, seeds that were implanted in the correct site but migrated outside the treatment site). 10 CFR 35.3045(b) states that a licensee shall report any event resulting from intervention of a patient or human research subject in which the administration of byproduct material or radiation from byproduct material results or will result in unintended permanent functional damage to an organ or a physiological system as determined by a physician.

Member of the Public – as defined in 10 CFR 20.1003, any individual except when that individual is receiving an occupational dose.

[I]**Neuralgia** – acute paroxysmal pain radiating along the course of one or more nerves usually without demonstrable changes in the nerve structure.

[I]**Neuroma** – a tumor or mass growing from a nerve and usually consisting of nerve fibers.

[II]**Neurosurgeon** – a physician trained in surgery of the nervous system and who specializes in surgery on the brain and other parts of the nervous system.

Non-stochastic Effect (Deterministic Effect) – as defined in 10 CFR 20.1003, health effects, the severity of which varies with the dose and for which a threshold is believed to exist. Radiation-induced cataract formation is an example of a non-stochastic effect.

Occupational Dose – as defined in 10 CFR 20.1003, the dose received by an individual in the course of employment in which the individual's assigned duties involve exposure to radiation or to radioactive material from licensed and unlicensed sources of radiation, whether in the possession of the licensee or other person. Occupational dose does not include doses received from background radiation, from any medical administration the individual has received, from exposure to individuals administered radioactive material and released under §35.75, from voluntary participation in medical research programs, or as a member of the public.

[I]**Paroxysm** – a sudden attack or spasm (as of a disease) or a sudden recurrence of symptoms or an intensification of existing symptoms.

Prescribed Dosage – as defined in 10 CFR 35.2, the specified activity or range of activity of unsealed byproduct material as documented (1) in a written directive or (2) in accordance with the directions of the authorized user for procedures performed pursuant to §§35.100 and 35.200.

Prescribed Dose – as defined in 10 CFR 35.2, (1) for gamma stereotactic radiosurgery, the total dose as documented in the written directive; (2) for teletherapy, the total dose and dose per fraction as documented in the written directive; (3) for manual brachytherapy, either the total source strength and exposure time or the total dose as documented in the written directive; or (4) for remote brachytherapy afterloaders, the total dose and dose per fraction as documented in the written directive.

"Prostate gland – a gland within the male reproductive system that is located just below the bladder.

Quality Factor (Q) – as defined in 10 CFR 20.1003, the modifying factor (listed in tables 1004(b).1 and 1004(b).2 of §20.1004) that is used to derive dose equivalent from absorbed dose.

Rad – as defined in 10 CFR 20.1004, the special unit of absorbed dose; one rad is equal to an absorbed dose of 100 ergs/gram or 0.01 Joule/kilogram (0.01 gray).

Radiation (ionizing radiation) – as defined in 10 CFR 20.1003, alpha particles, beta particles, gamma rays, x-rays, neutrons, high-speed electrons, high-speed protons, and other particles capable of producing ions; radiation, as used in 10 CFR Part 20, does not include nonionizing radiation such as radio waves or microwaves or visible, infrared, or ultraviolet light.

"Radiation Oncologist – a specialist in the use of radiation therapy as a treatment for cancer.

Radiation Safety Officer (RSO) – as defined in 10 CFR 35.2, an individual who (1) meets the requirements in §§35.50(a) or (c)(1) and 35.59; or (2) is identified as a radiation safety officer on (i) a specific medical use license issued by the Commission or Agreement State; or (ii) a medical use permit issued by a Commission master material licensee.

"Radiation Therapy (Radiotherapy) – in radiation therapy, high-energy rays are used to damage cancer cells and stop them from growing and dividing. A specialist in radiation therapy is called a radiation oncologist.

"Radiologist – a physician specialized in radiology, the branch of medicine that uses ionizing and nonionizing radiation for the diagnosis and treatment of disease.

Reactive Inspection – as defined in NRC Inspection Procedure 43003, "Reactive Inspections of Nuclear Vendors," an inspection performed for the purpose of obtaining additional information and/or verifying adequate corrective actions on reported problems or deficiencies.

'Rectosigmoid – the distal part of the sigmoid colon and the proximal part of the rectum

Rem – as defined in 10 CFR 20.1004, the special unit of any of the quantities expressed as dose equivalent; the dose equivalent in rems is equal to the absorbed dose in rads multiplied by the quality factor (1 rem = 0.01 sievert).

"Seminal vesicle – a structure in the male that is about 5 centimeters (2 inches) long and is located behind the bladder and above the prostate gland. The seminal vesicles contribute fluid to the ejaculate.

Shallow-dose Equivalent (H_S) – as defined in 10 CFR 20.1003, which applies to the external exposure of the skin of the whole body or the skin of an extremity, is taken as the dose equivalent at a tissue depth of 0.007 centimeter (7 mg/cm^2).

Sievert (Sv) – as defined in 10 CFR 20.1004, the internal system of unit of any of the quantities expressed as dose equivalent; the dose equivalent in sieverts is equal to the absorbed dose in grays multiplied by the quality factor (1 Sv = 100 rems).

Source Material – as defined in 10 CFR 70.4, source material as defined in section 11z. of the Act and in the regulations contained in Part 40 of this chapter; as defined in 10 CFR 40.4, means (1) uranium or thorium, or any combination thereof, in any physical or chemical form or (2) ores that contain by weight 1/20th of 1 percent (0.05 percent) or more of: (i) uranium, (ii) thorium, or (iii) any combination thereof. Source material does not include special nuclear material.

Special Nuclear Material – as defined in 10 CFR 70.4, (1) plutonium, uranium-233, uranium enriched in the isotope 233 or in the isotope 235, and any other material that the Commission, pursuant to the provisions of section 51 of the Act determines to be special nuclear material but does not include source material; or (2) any material artificially enriched by any of the foregoing but does not include source material.

Stereotactic Radiosurgery – as defined in 10 CFR 35.2, the use of external radiation in conjunction with a stereotactic guidance device to very precisely deliver a therapeutic dose to a tissue volume.

Stochastic Effects – as defined in 10 CFR 20.1003, health effects that occur randomly and for which the probability of the effect occurring, rather than its severity, is assumed to be a linear function of dose without threshold; hereditary effects and cancer incidence are examples of stochastic effects.

Therapeutic Dose – as defined in 10 CFR 35.2, a radiation dose delivered from a source containing byproduct material to a patient or human research subject for palliative or curative treatment.

Treatment Site – as defined in 10 CFR 35.2, the anatomical description of the tissue intended to receive a radiation dose as described in a written directive.

^{ll}Trigeminal Nerve – functions both as the chief nerve of sensation for the face and the motor nerve controlling the muscles of mastication (chewing). The trigeminal nerve is the fifth cranial nerve. The cranial nerves emerge from or enter the skull (the cranium) as opposed to the spinal nerves that emerge from the vertebral column. There are 12 cranial nerves.

^{l}Trigeminal Neuralgia – a very painful swelling (inflammation) of the nerve (trigeminal nerve) that delivers feeling to the face and "surface" of the eye.

^{ll}Urethra – the transport tube leading from the bladder to discharge urine outside the body.

Weighting Factor (w_T) – as defined in 10 CFR 20.1003 for an organ or tissue (T) the proportion of the risk of stochastic effects resulting from irradiation of that organ or tissue to the total risk of stochastic effects when the whole body is irradiated uniformly; weighting factors are listed in the table "Organ Dose Weighting Factors."

Whole Body – as defined in 10 CFR 20.1003, for purposes of external exposure, head, trunk (including male gonads), arms above the elbow, or legs above the knee.

Written Directive – as defined in 10 CFR 35.2, an authorized user's written order for the administration of byproduct material or radiation from byproduct material to a specific patient or human research subject, as specified in §35.40.

---

[i] These terms are not defined in 10 CFR, a management directive, an inspection procedure, or in a NRC policy statement. Rather, these terms are defined based upon definitions in Merriam-Webster's MedlinePlus Online Medical Dictionary. MedlinePlus is a service of the U.S. National Library of Medicine and the National Institutes of Health (http://www.nlm.nih.gov/medlineplus/mplusdictionary.html).

[ii] These terms are not defined in 10 CFR, a management directive, an inspection procedure, or in an NRC policy statement. Rather, these terms are defined based upon definitions in MedicineNet's Online MedTerms Medical Dictionary. MedicineNet is an online service part of WebMD (www.medterms.com).

[iii] This term is not defined in 10 CFR, a management directive, an inspection procedure, or in an NRC policy statement. Rather, these terms are defined based on the definitions in the online WebMD (www.webmd.com).

APPENDIX E
CONVERSION TABLE

Radioactivity and Ionizing Radiation

QUANTITY	FROM METRIC UNITS	TO NON-SI UNITS	DIVIDE BY
(Radionuclide) Activity	MBq	Curie (Ci)	37,000
	TBq	Ci	0.037
	GBq	Ci	37
Absorbed dose	Gy (gray)	rad	0.01
	cGy	rad	1.0
Dose equivalent	Sv (sievert)	rem	0.01
	cSv	rem	1.0
	mSv	rem	10
	mSv	mrem	0.01
	µSv	mrem	10

NRC FORM 335 (12-2010) NRCMD 3.7	U.S. NUCLEAR REGULATORY COMMISSION	1. REPORT NUMBER (Assigned by NRC, Add Vol., Supp., Rev., and Addendum Numbers, If any.)
	BIBLIOGRAPHIC DATA SHEET *(See instructions on the reverse)*	NUREG-0090, Vol. 33

2. TITLE AND SUBTITLE

Report to Congress on Abnormal Occurrences, Fiscal Year 2010

3. DATE REPORT PUBLISHED

MONTH	YEAR
June	2011

4. FIN OR GRANT NUMBER

5. AUTHOR(S)

6. TYPE OF REPORT

Annual

7. PERIOD COVERED *(Inclusive Dates)*

Fiscal Year 2010

8. PERFORMING ORGANIZATION - NAME AND ADDRESS *(If NRC, provide Division, Office or Region, U.S. Nuclear Regulatory Commission, and mailing address; if contractor, provide name and mailing address.)*

Division of System Analysis
Office of Nuclear Regulatory Research
U.S. Nuclear Regulatory Commission
Washington, DC 20555-001

9. SPONSORING ORGANIZATION - NAME AND ADDRESS *(If NRC, type "Same as above"; if contractor, provide NRC Division, Office or Region, U.S. Nuclear Regulatory Commission, and mailing address.)*

Same as 8, above

10. SUPPLEMENTARY NOTES

NRC Project Manager John J. Tomon

11. ABSTRACT *(200 words or less)*

Section 208 of the Energy Reorganization Act of 1974 identifies an abnormal occurrence (AO) as an unscheduled incident or event that the Nuclear Regulatory Commission (NRC) determines to be significant from the standpoint of public health or safety. The Federal Report Elimination and Sunset Act of 1995 requires that the AOs be reported to Congress on an annual basis. This report includes those events that the NRC has determined to be AOs during fiscal year 2010.

This report describes eight events at NRC-licensed facilities and seven events at Agreement State Licensed facilities that meet the criteria to be classified as AOs. In addition, this report provides an update to two events reported in fiscal year 2009 and four other events of interest.

12. KEY WORDS/DESCRIPTORS *(List words or phrases that will assist researchers in locating the report.)*

Exposure, Dose, Dosage, Medical Event, Fuel Facility, Nuclear Power Reactor

13. AVAILABILITY STATEMENT

unlimited

14. SECURITY CLASSIFICATION

(This Page)

unclassified

(This Report)

unclassified

15. NUMBER OF PAGES

16. PRICE

NRC FORM 335 (12-2010)

Printed
on recycled
paper

Federal Recycling Program

NUREG-0090
Vol. 33

REPORT TO CONGRESS ON ABNORMAL OCCURRENCES
FISCAL YEAR 2010

June 2011

UNITED STATES
NUCLEAR REGULATORY COMMISSION
WASHINGTON, DC 20555-0001

OFFICIAL BUSINESS